Wau Ecology Institute Pamphlet No. 2

Guide for Medical Orderlies: Medicinal Plants of Papua New Guinea

Plants Used for External Treatments
Buk bilong Dokta Boi long Bus

By

O. Gebia & K. Varpit

Guide for Medical Orderlies: Medicinal Plants of Papua New Guinea, Plants Used for External Treatments (Buk bilong Dokta Boi long Bus) (Wau Ecology Institute Pamphlet No. 2) by O. Gebia & K. Varpit

Originally published by the Wau Ecology Institute. Reprinted with Permission.

ISBN 978-9980-945-94-5

Copies are available from:

University of Papua New Guinea Press and Bookshop
PO Box 413
University PO, NCD
Papua New Guinea

FORWARD

Papua New Guinea National Government spent millions of Kina annually on imported drugs for various diseases, starting from simple sore and cough treatments to major killer diseases like Malaria and Cancer.

It is interesting to note that many of the drugs used have been abstracted from the living organisms such as plants. In Papua New Guinea before the arrival of modern drugs many of our people has resorted to herbal medicine. Today still many of our people do not have easy access to modern drugs, because the nearest aidpost maybe many miles away from the village. Some times the aidpost is not properly equipped to attend to the sick patient that has walked many kilometres.

Papua New Guinea with over three hundred species of Plants do certainly have potential for drug development. I believe this booklet will not only be a guide to medical orderlies, but do hope it will be an eye opener for our Medical Professions, the Politicians and the public at large that we need to put financial resources into developing our own drugs.

The German Government and EZE deserve special thanks for supporting this publication, which I believe will benefit many of our potential users.

Harry Sakulas, MSc
Director

INTRODUCTION IN ENGLISH

This pamphlet has been prepared primarily for the Aid Post Orderlies (APO's) in Papua New Guinea. It is based especially on plants that are used for external treatments by the local people from various parts of the country where the Staff of Wau Ecology Institute had conducted surveys on medicinal plant uses.

This pamphlet was produced mainly to give basic knowledge to the APO's on the external uses of medicinal plants and to make them aware that there are other local means of medical application to external treatments which they should inform the communities when the application of modern medicine is not so effective. This could help the people who are in need of Health services into their rural areas.

In other words the people will be able to realize that they own the much cheap and appropriate medicine in their surrounding environment. Some of our local medicine are much effective than the modern medicine and drugs.

The medicinal plant information on the uses especially on external treatments were extracted from the reports that were produced by the Wau Ecology Institute's Medicinal Plant Project and also from the Part I of the Morobe Province Medicinal Plant Handbook.

The pamphlet presents a list of plant species along with their botanical description, their medicinal use and the methods of application and precaution. The text have also been translated to Pidgin to make it easier and understandable for the Melanesian Pidgin speakers. The photographs provided may be a help for identification of the plants together with their brief descriptions.

This pamphlet can be a helpful guide to the APO's, the community and to anybody interested in the traditional application of the plants. I believe that the Wau Ecology Institute will produce several of this type of leaflets in the near future.

INTRODUCTION IN PIDGIN

Mipela i kamapim dispela liklik buk bilong helpim ol Medikol Odali long Papua New Guinea. Dispela buk i stori moa long diwai, rop na gras ol tumbuna bilong yumi isave usim long bipo long oraitim ol sik or bagarap isave kamap autsait long bodi bilong ol man na meri long plenti hap long Papua Niugini. Mipela ikisim halivim long ol bikman long peles we mipela ibin kerim aut dispela projek. Ol bikman husait igat save long usim ol tumbuna marasin bilong bipo.

Dispela liklik buk mipela i wokim em bilong givim tingting na save long ol Medikol Odali long bus marasin we ol pipol isave usim long oraitim ol sik or bagarap isave kamap autsait long bodi bilong ol man na meri. Na tu ol Medikol Odali iken save olsem igat arapela marasin we ol iken toksave long ol pipol taim ol marasin bilong ol waitman ino istap. Dispela kain save ken givim gutpela halivim long ol man meri husait istap lon we long hausik.

Long mekim klia, ol pipol bai gat klia tingting na save long ol kain marasin we emi fri na isi long panim insait long asples bilong yumi yet. Sampela marasin bilong yumi igat kik winim bilong waitman.

Ol toktok bilong marasin insait long dispela liklik buk na long ol wok bilong en (arasait long skin) ibin kam aut long ripot we Wau Ecology Institute Medisinel Plent Projek ibin kamapim na tu long part wan (1) bilong Morobe Provinsel medisinel plent henbuk.

Dispela liklik buk igivim nem bilong ol wanwan diwai wantaim tok klia na as bilong diwai, wok bilong marasin, na wei bilong usim na tok lukaut. Ol toktok insait long buk tu mipela ibin tanim long tok Pidgin bai mekim isi na klia long ol man meri iritim. Ol piksa insait tu iken halivim long tok klia na luk save long ol diwai.

Dispela liklik buk iken halivim na stiaim ol Medikol Odali, ol komuniti na husait man igat laik long usim tumbuna marasin. Mi bilip olsem Wau Ecology Institute bai kamapim plenti bilong ol dispela kain liklik buk long bihain taim.

Scientific Name: *Carica papaya* L.
Common Name : Popo (Keregia - Finschafen - Morobe)
Description : Its fruit is juicy and edible.

Distribution : Widely found in the country.
Application : Treatment for ringworm and grille.
How to use : Usim olsem wanem.

1. Collect several green fruit from the tree.
(Kisim sampela grin prut long diwai).

2. Cut the fruit.
(Katim prut).

3. Collect milky juice from a dish/bowl.
(Kisim susu bilong popo long dis).

4. Apply directly to the infected area.
(Putim susu bilong popo igo long hap bodi igat bagarap).

Precaution : Avoid getting the juice on to your eyes.

Scientific name: *Amomum aculeatum* **Roxb.**
Local name : Obo (Maprik - E.S.P.)
Description : A plant that doesn't grow too high. Its stem is not too hard. a herb, 2-4m. Fruits of these herb are edible which is known as "gorgor" in pidgin.

Distribution : Widely distributed in the country.
Application : Treatment for fresh cuts.
How to use : (Usim olsem wanem).

1. Collect the stem.
   (Katim namel bilong en).

2. Heat it in the open fire.
   (Hatim long paia).

3. Crush it and collect saps in the bottle.
   (Brukim na kisim wara bilong en).

4. Apply it directly to the fresh cut to stop bleeding.
   (Putim stret igo long ples nuipela sua istap long en).

Scientific name: *Alstonia scholaris* **R. Br.**
Local name : Boyom
Description : A tree, 30-50m tall. The bark, stem and leaves exudates while sap. Common in lowland and lower montane rainforest.

Distribution : Raikos - Madang - Lumi West Sepik
Application : Treatment for boil
How to use : (Usim olsem wanem).

1. Break off the leave from the stem.
(Brukim lip long diwai).

2. Collect the sap.
(Kisim susu bilong diwai).

3. Apply to the newly formed boil as a poultice.
(Putim igo long hap bodi buk (boil) i kamap).

Scientific name: *Bidens pilosa* **L.**
Local name : Kwipo
Description : A plant. It can be domestically grown. An erect annual herb. A weed of roadside, forest clearings and cultivated areas.

Distribution : Watut - Morobe Province. It's also found in other parts of the country.
Application : Treatment for Boil.
How to use : (Usim olsem wanem).

1. Collect the flowers.
   (Kisim ol plaua bilong em).

2. Gently press directly onto the boil.
   (Isi isi na putim stret antap long buk (boil).

3. It enables the collection of apss in the tissue.
   (Em i helpim long rausim susu bilong buk (boil) insait.

Scientific name: *Casia alata* L.
Local name : Gala (Kerengia - Morobe)
: Kubosi (Mingming - Madang)
: Amber (Lorein - Maprik E.S.P.)
Description : A tree commonly found near the river bank. It can not grown too high. A shrub, to 3m tall. A plant of waste land.

Distribution : Widely distributed in the country.
Application : Treatment for grille.
How to use : (Usim olsem wanem).

1. Collect the leaves.
   (Kisim lip bilong diwai).

2. Crush the leaves.
   (Brukim lip igo malumalum).

3. Apply directly to the infected area.
   (Putim igo long hap igat sik long en).

4. Leave it on your skin for 3-4 hours when wash it off.
   (Lusim istap long skin inap long 3-4 awa na bihain wasim).

5. Repeat if necessary.

Scientific name: *Cordyline terminalis* **Kunth.**
Local name : Kawa (Nobnob - Madang)
Description : A shrub. Mostly grown around residential areas and cultivated areas.

Distribution : Found in Madang.
Application : Treatment for fresh knife cuts.
How to use : (Usim olsem wanem)

1. Collect several leaves according to the size of wound.
   (Kisim lip inap long sua).

2. Crush the leaves in a bowl.
   (Brukim ol lip - memeim igo liklik long boul (dis).

3. Apply the saps directly to the wound.
   (Putim wara bilong en igo stret long sua).

4. If bandage is required, use it as a bandage too.
   (Yu laikim banis, usim olsem banis tu).

Scientific name: *Cycas circinalis* **L.**
Local name : Tupeh (Sosoningko - Morobe)
: Tuwon (Lalok - Madang)
: Okineng (Lorein - Maprik - ESP)
Description : It is like a palm tree. But doesn't grow too high. Normally found in the grassland vegetation.

Distribution : Mostly found in the savannah grassland.
Application : Treatment for tropical ulcers.
How to use : (Usim olsem wanem).

1. Collect seeds.
   (Kisim ol sids (pikinini) bilong en).

2. Cut it open.
   (Katim na opim).

3. Scrap the kernel onto the sores.
   (Sigirapim mit igo long sua).

Scientific name: *Erigeron sumatrensis* **Retz.**
Local name : Guhehane (Buang - Morobe Province)
Description : It is a plant. The leaves are bit sticky mostly grown. Erect annual herb. A weed of roadsides, abandoned garden areas.

Distribution : Buang - Morobe Province
Application : Treatment for fungal infection and grille.
How to use : (Usim olsem wanem).

1. Collect young leaves.
   (Kisim yangpela lip)

2. Rub it into an infected area.
   (Rabim long hap bodi igat sik long en).

Scientific name: *Homalomena aromatica* **(Sima.) Schott.**
Local name : Waurjag
Description : A herb it is like a chinese taro. Crushed leaves and stems of this plant produce pleasant smell.

Distribution : Found in Gogol - Madang - Also found in rest of the country.
Application : Treatment for boil.
How to use : (Usim olsem wanem)

1. Collect leaves.
   (Kisim lip).

2. Put it on the boil.
   (Putim stret long boil (buk).

3. Use it as poultice until it bursts.
   (Karamapim stap inap boil (buk) i bruk.

Scientific name: *Honstaedtia lycostoma* **K. Schum.**
Local name : Hiru (Gogol - Madang)
Description : A herb, 2-4m high. Occuring sometimes in successional forest vegetation and in primary forest.

Distribution : Found in Gogol valley in Madang. It's also commonly found in PNG.
Application : Treatment for old and new wounds.
How to use : (Usim olsem wanem).

1. Heat the stem over the fire.
   (Katim gorgor na hatim long paia).

2. Squeeze the sap onto the cut or an old wound directly.

Scientific name: *Merremia peltata* **(L.) Merr.**
Local name : Wafigo (Finschaffen - Morobe)
Description : A creeper and climber (vine). Common along edges of primary and secondary forest, from sea level to 500m.

Distribution : Found in Finschaffen - Morobe.
Application : Treatment for axe and knife wound.
How to use : (Usim olsem wanem).

1. Cut the stem.
   (Katim namel bilong en).

2. Collect saps in a container.
   (Pulimapim wara bilong en long botol samting).

3. Apply directly to the wound.
   (Putim igo stret long sua).

Scientific name: *Pipturus argenteus* **(Forst.) Wedd.**
Local name : Lul (Buang - Morobe)
: Zil (Gogol - Madang)
Description : A shrub to a small tree. Common in secondary forest.

Distribution : Found in Buang and Gogol area of Morobe and Madang.
Application : Treatment for extracting broken spear stuck in human flesh and also for scabies.
How to use : (Usim olsem wanem).

1. Collect the bark.
   (Kisim skin bilong en).

2. Scrap it into a container.
   (Sigirapim skin bilong en igo long botol).

3. Gently apply to the speared area and extract the spear.
   (Putim isi isi igo long hap bodi spia i pas long en na rausim spia).

4. Mix the powdered bark with water.
   (Mixim pauda skin wantaim wara).

5. Bath with it to treat scabies.
   (Waswas wantaim na rausim scabies (kaskas).

Scientific name: *Planchonia papuana* **R. Kunth.**
Locally used by: Maprik people, ESP
Description : A large tree to 40m tall. In lowland rainforest especially along creeks and river sides.

Distribution : Found in Saulik (Maprik) East Sepik.
Application : Treatment for scabies.
How to use : (Usim olsem wanem).

1. Collect the bark.
   (Kisim skin bilong en).

2. Crush it.
   (Brukim igo malumalum).

3. Mix with water.
   (Mixim wantaim wara).

4. Bath with it.
   (Waswas wantaim en).

Scientific name: *Plecthranthus scutellarioides* **(L.) R. Br.**
Local name : Miango (Nauti - Morobe Province)
Description : A herb - 1 m tall. Common in gardens and residential areas.

Distribution : Found in Nauti - Morobe Province. It's also found elsewhere in the country.
Application : Treatments for axe, knife or spear wounds.
How to use : (Usim olsem wanem).

1. Collect the stems.
   (Kisim namel bilong en).

2. Chew it.
   (Kaikaim long maus).

3. Spit the saps onto the wounds.
   (Spetim wara igo long sua).

Scientific name: *Rhododendron macgregoriae* L.
Local name : Kanoma (Aseki - Menyamya)
Description : A shrub to a small tree (5m tall). At forest grassland margins, in successional regrowth areas. Occurs in the montane forest areas.

Distribution : Found in Aseki - Morobe Province
Application : Treatment for tropical ulcer.
How to use : (Usim olsem wanem).

1. Collect leaves.
   (Kisim lip)

2. Crush to pieces
   (Brukim igo malumalum).

3. Mix with few drops of water.
   (Mixim wantaim liklik wara).

4. Apply directly to the sores.
   (Putim igo stret long sua).

ACKNOWLEDGEMENT IN ENGLISH

There are many who have contributed to this pamphlet in one way or another. Firstly we wish to thank all of the village people encountered in our surveys over the years who willingly shared their important and valuable knowledge.

We gratefully acknowledge our funding organization, EZE through Wau Ecology Institute for providing logistics and financial support during the on going of this project.

Furthermore we would like to thank the staff of the National Herbarium, Lae, for the identification of numerous plant specimens.

We are grateful to Mr. Joseph Geparo for his assistance in rearranging and formatting the text and also translating the English version to Melanesian Pidgin version.

Finally we would like to thank the Director of Wau Ecology Institute, Mr. Harry Sakulas for his support and interest during the on going of this project **and also editing of the manuscript.**

ACKNOWLEDGEMENT IN PIDGIN

Igat plenti man meri husait i bungim tingting long kamapim dispela liklik buk. Pastaim tru, mipela laik tok tenkyu igo long ol man meri bilong peles husait i halivim long givim gutpela tingting na save bilong tumbuna marasin bilong ol, long dispela taim long wok panim aut bilong mipela.

Mipela tu i givim bikpela tok hamamas igo long EZE, long Germany, long givim mani ikam long Wau Ecology Institute long halivim na sapotim mipela long dispela projek.

Igo mo yet tu, mipela laik tok tenkyu igo long ol wok manmeri bilong National Herbarium, Lae, long painim aut nem bilong dispela ol kainkain diwai, rop na gras ol tumbuna isave usim bipo.

Mipela tu tok hamamas igo long Mr. Joseph Geparo long halivim long stretim na wokim kamap dispela buk na tu tanim tok waitman igo long tok pidgin.

Long pinisim mipela laik tok tenkyu long Dairekta bilong Wau Ecology Institute, Mr. Harry Sakulas husait igat bikpela laik na sapotim mipela long mekim dispela buk i kamap olsem.

REFERENCES

Ayensu, E.S. 1979. Plants for Medicinal Uses with Special Reference to Arid Zones. In: *Arid Land Plant Resources Goodin J.R. and D.K. Northington (eds.), Smithsonian Institute, Washington, D.C.*

Cribb, A.A. 1981. Wild Medicine of Australia. *Collins, Sydney, Australia.*

Ladion, H.G. 1985. Healing Wonders of Herbs, Guide to Effective Use of Medicinal Plants. *Manila, Philippines.*

Perry, L.N. 1980. Medicinal Plants of East and South East Asia. *M.I.T. Press, Cambridge, Mass, USA.*

Woodley, E. 1991. Medicinal Plants of Papua New Guinea. Part I, Morobe Province. Wau Ecology Institute Handbook No. 11. *Verlog Josef Margraf Scientific Books.*

www.ingramcontent.com/pod-product-compliance
Lightning Source LLC
Chambersburg PA
CBHW061307300425
25819CB00006B/24